Your Guide to Home Electrical Wiring, Outlet and Switch Installs

DIY Instructions for Circuit Maps,
Running New Wires, Installing Fixtures,
Replacing Old Outlets and Switches
Safely to Code

Savvy Quick Fix Joel

Table of Contents

Introduction

Did you know that the average household spends nearly $2,000 each year on electricity? That's not just a number; it's hard-earned money slipping through our fingers and into the power grid. As the world becomes more connected, our reliance on electricity grows, and so does the burden on our wallets. If you've ever opened your monthly utility bill with a sense of dread or wondered how you could cut costs without sacrificing comfort, you're not alone.

We understand the frustrations and concerns that come with managing home energy expenses. The fear of the unknown, the uncertainty around DIY electrical work, and the worry about making costly mistakes can be paralyzing. However, there's a beacon of hope amidst the tangled wires and outdated fixtures of your home. Welcome to "Your Guide to Home Electrical Wiring, Outlet and Switch Installs," where we empathize with your struggles and offer you a roadmap to not just save money but also gain control over your home's energy destiny.

In the chapters that follow, we'll delve into the fundamentals of electrical systems, providing you with the knowledge to decipher the mysteries behind your utility bills. Imagine the satisfaction of mapping out your home's circuits, the confidence of running new wires with ease, and the empowerment to install fixtures and replace outlets and switches without relying on expensive professionals. This book is your passport to a realm where you control your energy destiny, and each project you tackle brings you one step closer to a more efficient, cost-effective, and environmentally friendly home.

By the time you reach the last page, you won't just have conquered your fear of electrical work; you'll have transformed your relationship with energy consumption. Your efforts will translate into tangible benefits – lower bills, a more sustainable lifestyle, and the pride that comes with mastering a skill that enhances both your home and the environment.

Don't let the complexities of home electrical systems keep you in the dark. Join us on this enlightening journey, and let's empower you to take charge of your energy future. It's time to illuminate your home, your savings, and your impact on the world. Are you ready to flip the switch?

Chapter 1

Understanding Electrical Basics

Voltage, Current, and Resistance

Welcome to the electrifying world of home wiring! In this chapter, we'll unravel the mysteries behind the three key players in the electrical game: voltage, current, and resistance.

1. Voltage: The Electric Push

Voltage is the force pushing electricity through wires, think of it as the pressure in a water pipe. It's what makes your lights shine and your appliances hum. High voltage means a powerful push, and low voltage means a gentler nudge. Voltage is measured in volts (V).

Key Takeaway: Voltage is the driving force of electrical flow, the push that keeps your home powered.

2. Current: The Flow of Electricity

Current is the actual flow of electricity along a conductor, like the water flowing through a pipe. It's measured in amperes (A). Your devices and appliances draw a certain amount of current to function.

Key Takeaway: Current is the movement of electricity, and different devices require different amounts to operate.

3. Resistance: The Speed Bump
Resistance is what opposes the flow of electricity. It's like the speed bumps on your street, it slows things down. The unit of resistance is the ohm (Ω). Different materials have different resistances.

Key Takeaway: Resistance is the hindrance to the flow of electricity, and it varies depending on the material.

Recap: Voltage is the push, current is the flow, and resistance is the speed bump.

Now, let's put it into perspective. Imagine a water hose (the wire) connected to a faucet (voltage). When you turn on the faucet, water (current) flows through the hose. If you squeeze the hose (increase resistance), the water flow decreases. Simple, right?

Why does this matter to you?
Understanding these basics empowers you to troubleshoot issues, choose the right wires for your projects, and ensure your home's electrical system runs smoothly. Now that you've got the keys to the electrical

kingdom, let's move on to the next chapter, where we'll explore the types of circuits awaiting your mastery.

Key Takeaway: Voltage, current, and resistance are the fundamental elements of electrical systems. Just like knowing how water flows through a hose helps you manage your garden, understanding these concepts empowers you to manage the flow of electricity in your home. Get ready to shine a light on your newfound electrical knowledge!

Types of Circuits

Welcome to the fascinating realm of circuits, the intricate pathways that bring electricity to life in your home. Understanding the different types of circuits is like having a versatile toolbox, each type has its purpose, and knowing when to use them ensures your electrical projects run smoothly. Let's dive in!

1. Series Circuits: The Chain Reaction

Think of series circuits as a string of fairy lights, one goes out, they all go out. In a series circuit, devices are connected end-to-end, forming a single pathway for current flow. The catch? If one device fails, the entire circuit breaks, leaving everything downstream powerless.

Why it Matters: Series circuits are simple but come with a "domino effect" a failure in one component affects the entire circuit.

2. Parallel Circuits: The Power Network

Imagine your home's electrical outlets, that's a parallel circuit. In this setup, each device has its own path to the power source. If one device fails, the others continue to function independently. Parallel circuits provide redundancy and are a common choice for wiring outlets and lights in your home.

Why it Matters: Parallel circuits offer reliability, a failure in one device doesn't disrupt the operation of others.

3. Combination Circuits: The Hybrid Approach

Combination circuits take the best of both worlds, series and parallel. They're like the Swiss Army knife of electrical setups, offering redundancy and a continuous flow of current. Larger electrical systems often use combination circuits for increased flexibility and reliability.

Why it Matters: Combination circuits provide a balance between the simplicity of series circuits and the reliability of parallel circuits.

4. AC (Alternating Current) and DC (Direct Current) Circuits: The Directional Dilemma

AC and DC refer to the direction in which current flows. AC alternates direction periodically, the type of current you find in your home's outlets. DC flows consistently in one direction, commonly used in batteries. Knowing which type your devices need ensures compatibility.

Why it Matters: Different devices and systems require either AC or DC, understanding the distinction is crucial for proper functioning.

As you embark on your journey of DIY electrical projects, grasp these circuit types like a seasoned electrician. Whether you're lighting up a room, connecting appliances, or troubleshooting issues, choosing the right circuit type is the secret sauce for success. Now armed with this knowledge, go forth and wire with confidence!

Key Takeaway: Series circuits form a chain, parallel circuits create a network, combination circuits offer the best of both, and AC/DC circuits cater to specific device needs. Mastering these circuit types empowers you to handle any electrical project that comes your way. Get ready to illuminate your home with newfound expertise!

Safety Precautions and Regulations

Now that we've laid the groundwork for understanding electrical basics and circuit types, let's delve into a critical aspect of any electrical project – safety. Electricity is a powerful force, and ensuring your well-being and the integrity of your home's electrical system requires strict adherence to safety precautions and regulations.

1. Personal Protective Equipment (PPE): Your Armor

When venturing into the world of home electrical projects, consider your PPE as your armor. Equip yourself with essentials like insulated gloves, safety glasses, and non-conductive footwear. These items serve as your first line of defense against potential electrical hazards.

Why it Matters: PPE protects you from electric shocks, burns, and other injuries during the project.

2. Power Off: The Golden Rule

Before laying a hand on any electrical component, always ensure the power is turned off. Locate and switch off the circuit breaker or disconnect the power source. Use lockout/tagout procedures if possible, preventing accidental re-energizing by others.

Why it Matters: Cutting power minimizes the risk of electric shock and other accidents.

3. Inspect Tools and Equipment: Your Trusted Allies

Faulty tools can turn a simple project into a hazardous situation. Before starting, inspect all tools and equipment for damage. Ensure that cords are intact, insulation is secure, and any metal parts are free from defects.

Why it Matters: Well-maintained tools reduce the risk of accidents and ensure efficient work.

4. Know Your Limits: DIY Wisely

While DIY projects can be empowering, it's crucial to recognize your limits. If a project involves complex wiring or unfamiliar systems, consider consulting a professional electrician. Your safety is paramount.

Why it Matters: Avoiding projects beyond your expertise prevents accidents and ensures quality work.

5. Adherence to Codes and Regulations: The Legal Guardian

Familiarize yourself with local electrical codes and regulations. These standards are in place for a reason – to safeguard you, your home, and those who may inhabit it. Ignoring them can lead to serious consequences.

Why it Matters: Compliance with codes ensures the safety and legality of your electrical work.

6. Grounding and Bonding: The Stability Pillars

Proper grounding and bonding are essential for a safe electrical system. Grounding provides a path for electrical faults to dissipate harmlessly, while bonding ensures electrical continuity. Both are critical for preventing shocks and fires.

Why it Matters: Grounding and bonding create a stable electrical environment, reducing the risk of accidents.

Recap: Safety precautions, including PPE, powering off, tool inspection, and knowing your limits, are non-negotiable. Adhering to local codes and regulations, and understanding the importance of grounding and bonding, ensures not only your safety but also the integrity of your electrical projects.

As we continue this journey, remember that safety is your constant companion. By embracing these precautions and regulations, you're not just ensuring a secure workspace, you're paving the way for successful and worry-free DIY electrical endeavors.

Chapter 2

Tools and Materials

Essential Tools for DIY Electrical Work

In the world of DIY electrical projects, having the right tools is akin to wielding a magic wand. It's the key to unlocking your potential and transforming your home into a well-wired haven. Let's explore the essential tools that will make your electrical endeavors not just possible, but efficient and safe.

1. Voltage Tester: The Electric Detective

First things first, always ensure the power is off before working on any electrical project. A voltage tester is your trusty sidekick in this quest. It helps confirm that wires are de-energized, preventing unexpected shocks.

Why it Matters: Ensures your safety by confirming the absence of electrical current before you start working.

2. Screwdrivers: The Versatile Allies

A good set of screwdrivers, both flathead and Phillips, is the backbone of any toolkit. From securing outlets to tightening wire connections, these versatile tools are indispensable.

Why it Matters: Allows you to handle a variety of tasks, ensuring a secure and stable electrical system.

3. Wire Strippers: Precision Unveiling
Stripping wires with precision is a crucial skill in electrical work. Wire strippers are designed for this task, allowing you to remove insulation without damaging the conductor.

Why it Matters: Ensures proper electrical connections by exposing the right amount of wire for termination.

4. Pliers: The Grip Masters
Whether you're bending wires, pulling cables, or crimping connectors, a good pair of pliers is a must. Needle-nose pliers are particularly handy for intricate work.

Why it Matters: Provides the necessary grip and control for various tasks in electrical installations.

5. Tape Measure: Precision in Proportions
Accuracy is paramount in electrical work, and a tape measure is your tool for achieving it. From spacing

outlets to measuring wire lengths, it ensures everything fits seamlessly.

Why it Matters: Allows you to plan and execute projects with precision, minimizing errors.

6. Wire Nut Connectors: The Junction Magicians
Connecting wires securely is a common task in electrical projects. Wire nut connectors simplify this process by providing a reliable and insulated junction.

Why it Matters: Facilitates safe and efficient connections, ensuring the integrity of your electrical system.

7. Fish Tape: Navigating Tight Spaces
Running wires through walls or conduits requires finesse. Fish tape is a flexible tool that aids in guiding wires through confined spaces, making seemingly challenging tasks more manageable.

Why it Matters: Simplifies the process of routing wires in walls or conduits, saving time and effort.

8. Circuit Tester: Unraveling Mysteries
Troubleshooting electrical issues is part of the DIY journey. A circuit tester helps identify faults, ensuring that your electrical circuits are in top-notch condition.

Why it Matters: Enables you to diagnose and resolve electrical problems, maintaining a functional system.

Recap: Essential tools such as voltage testers, screwdrivers, wire strippers, pliers, tape measures, wire nut connectors, fish tape, and circuit testers form the arsenal of a DIY electrician. Each tool serves a specific purpose, contributing to the efficiency, safety, and success of your electrical projects.

With these tools in hand, you're not just a DIY enthusiast, you're a DIY electrician, ready to tackle any wiring challenge that comes your way. Equip yourself with these essentials, and let the sparks of creativity and safety fly as you illuminate your home with your newfound electrical prowess.

Choosing the Right Wiring Materials

In the intricate dance of DIY electrical work, the choice of wiring materials is a crucial step. Selecting the right wires ensures not only the efficiency of your electrical system but also the safety of your home. Let's unravel the mysteries of wiring materials and empower you to make informed decisions for your projects.

1. Types of Electrical Wires

a. Non-Metallic (NM) Cable (Romex): The All-Rounder

Non-metallic cables are a popular choice for residential wiring. Often referred to as Romex, they consist of insulated conductors bundled together in a flexible plastic sheath. Suitable for various applications, Romex is easy to work with and versatile.

b. Armored Cable (AC): Robust Protection

Armored cables feature a metal sheath providing extra protection, making them suitable for areas where the wiring may be exposed to physical damage. This robust option is commonly used in industrial and commercial settings.

c. Conduit Wiring: The Flexible Shield

Conduit is a protective tube that encases individual wires or cables. It adds an extra layer of protection against damage and is often used in exposed or hazardous locations. Conduit wiring offers flexibility and customization.

2. Wire Gauge: Size Matters

a. Thick vs. Thin: Gauge Decisions

The gauge of a wire refers to its thickness. Thicker wires (lower gauge) can carry more current over longer distances without overheating. Understanding the electrical load and distance helps determine the appropriate wire gauge for your project.

b. Ampacity: Matching Current Capacity

Each wire gauge has a specific ampacity, indicating its safe current-carrying capacity. Ensuring that the wire's ampacity matches or exceeds the current requirements of your devices prevents overheating and potential hazards.

3. Insulation Material: Safeguarding Conductors

a. PVC (Polyvinyl Chloride): Common and Affordable

PVC is a widely used insulation material due to its affordability and versatility. It provides good protection against moisture and physical damage, making it suitable for various applications.

b. THHN/THWN: Heat and Moisture Resistance

Thermoplastic High Heat-Resistant Nylon (THHN) and Thermoplastic Heat and Water-Resistant Nylon (THWN) are types of wire insulation designed for enhanced heat and moisture resistance. They are commonly used in conduit wiring.

4. Consider Environmental Factors

a. Indoor vs. Outdoor Wiring

Outdoor wiring requires materials that can withstand exposure to the elements. UV-resistant jackets and waterproof insulation are essential for outdoor applications.

b. Wet vs. Dry Locations

In wet locations, such as bathrooms or kitchens, using wiring materials with moisture-resistant features is crucial. This prevents corrosion and reduces the risk of electrical issues.

Recap: Choosing the right wiring materials involves understanding the types of electrical wires, selecting the appropriate wire gauge, considering insulation materials, and accounting for environmental factors. Each decision contributes to the safety, efficiency, and longevity of your electrical system.

As you embark on your DIY electrical projects, the materials you choose will form the backbone of your installations. By mastering the art of selecting the right wires, you're not just ensuring the success of your projects, you're creating a safe and reliable electrical infrastructure within your home. Let's wire up your space with confidence and precision.

Chapter 3

Planning Your Electrical Project

Creating a Circuit Map

Embarking on a DIY electrical project without a plan is like setting sail without a map. In this chapter, we'll explore the crucial step of creating a circuit map. This visual roadmap not only guides your hands but also safeguards your journey through the intricate wiring of your home.

1. Assess Your Electrical Needs

a. Identify Electrical Devices and Fixtures:
Begin by listing all the electrical devices and fixtures you plan to install or modify. This could include outlets, light fixtures, switches, and appliances.

b. Determine Power Requirements:
Understand the power requirements of each device. Note the voltage, current, and wattage specifications. This information helps you choose the right wires and ensures your circuits can handle the load.

2. Sketch Your Home Layout

a. Blueprint Your Home:

Obtain or create a blueprint of your home. This provides an overview of the layout, helping you visualize the locations of walls, rooms, and key electrical components.

b. Mark Device Locations:

On the blueprint, mark the intended locations for outlets, switches, and fixtures. Consider factors like furniture placement and room usage to optimize the positioning of electrical elements.

3. Plan Circuit Layouts

a. Group Devices by Function:

Group devices with similar functions together. For example, lights in a room, outlets in a kitchen, or appliances in a laundry area. This organization streamlines the wiring process.

b. Avoid Overloading Circuits:

Distribute the electrical load evenly across circuits to prevent overloading. Balance the number of devices on each circuit to ensure safe and efficient operation.

4. Create a Visual Circuit Map

a. Use Symbols and Labels:

Represent electrical components with standardized symbols. Outlets, switches, and lights have specific symbols that make your circuit map clear and easy to follow. Label each element for clarity.

b. Indicate Wiring Paths:

Draw lines to illustrate the wiring paths between devices. Differentiate between different types of wires, such as those for lighting circuits, outlets, or appliances.

5. Incorporate Safety Measures

a. Highlight Emergency Shut-offs:

Identify emergency shut-off locations on your circuit map. This could include circuit breaker panels or shut-off switches for specific circuits. Ensuring easy access to shut-offs enhances safety during maintenance or emergencies.

b. Note Conduit and Wiring Routes:

If you're using conduits, mark the routes on your map. Understanding the path of wiring simplifies troubleshooting and future modifications.

Recap: Creating a circuit map involves assessing your electrical needs, sketching your home layout, planning circuit layouts, and visually mapping the circuits. This comprehensive plan acts as your guiding compass through the intricacies of home electrical wiring.

As you finalize your circuit map, remember that this document is your blueprint for success. It not only ensures a systematic and organized approach to your electrical project but also serves as a valuable reference for future maintenance or renovations. Armed with a well-crafted circuit map, you're ready to bring your electrical vision to life. Let's wire up your home with precision and confidence.

Calculating Electrical Load

Understanding and calculating the electrical load of your home is akin to knowing how much weight a bridge can bear. It ensures the safe and efficient distribution of electrical power, preventing overloads and potential hazards. In this chapter, we'll unravel the process of calculating electrical load, empowering you to navigate the currents of your home's electrical system with confidence.

1. Define Electrical Load

a. What is Electrical Load?
   Electrical load refers to the total power consumed by all connected devices and appliances in a given electrical circuit or system. It's measured in watts (W) or kilowatts (kW).

b. Types of Loads:
i. Resistive Load: Devices like incandescent bulbs and electric heaters primarily produce heat as a byproduct.

ii. Inductive Load: Devices with motors, such as refrigerators and air conditioners, have an inductive load due to their moving parts.

2. Identify Devices and Appliances

a. List all Devices:
Create a comprehensive list of all devices and appliances connected to a specific electrical circuit. Include lighting fixtures, outlets, and any devices powered by the circuit.

b. Determine Power Ratings:
Check the power rating (in watts or kilowatts) of each device. This information is usually found on a label or in the device's manual.

3. Calculate Power Consumption

a. Wattage Calculation:
For devices with a constant power draw, the calculation is straightforward. Multiply the device's power rating (in watts) by the number of hours it operates in a day.

$$\text{Power Consumption (Wh)} = \text{Power Rating (W)} \times \text{Operating Hours per Day}$$

b. Inductive Load Adjustment:
Inductive loads, like motors, may have a higher initial power requirement (starting watts) before settling into a lower operational state (running watts). Account for this difference when calculating power consumption.

4. Determine Circuit and Overall Load

a. Sum Individual Loads:
Add up the power consumption values for all devices connected to a specific circuit. This gives you the load for that circuit.

b. Check Circuit Capacity:
Compare the calculated load to the circuit's capacity. Circuits are typically rated in amperes (A) or kilowatts. Ensure that the calculated load does not exceed the circuit's capacity to prevent overloads.

5. Future-Proofing and Redundancy

a. Consider Future Additions:
Anticipate any future additions or changes to the electrical system. Leave some room for growth when calculating the load to accommodate potential new devices.

b. Implement Redundancy Measures:
Distribute loads across multiple circuits to avoid relying heavily on a single circuit. This provides redundancy and prevents overloading.

Recap: Calculating electrical load involves defining electrical load, identifying devices and appliances, calculating power consumption, determining circuit and overall load, and considering future-proofing measures. This process ensures the safe and efficient operation of your electrical system.

As you embark on calculating the electrical load for your home, envision it as orchestrating a symphony of power. By understanding and managing the load, you're not just preventing potential electrical issues – you're ensuring a harmonious and efficient distribution of energy throughout your home. Let's bring balance to your electrical system and keep the currents flowing smoothly.

Obtaining Necessary Permits

Embarking on a home electrical project without the required permits is like setting sail without a compass. Permits are not just bureaucratic hurdles; they are crucial safeguards that ensure your electrical work complies with safety standards and local regulations. In this chapter, we'll navigate the process of obtaining necessary permits, guiding you through the essential steps to keep your project on the right course.

1. Understand the Importance of Permits

a. Legal Compliance:

Permits are not just paperwork; they are a legal requirement. Local building codes mandate permits for electrical work to ensure that installations meet safety standards and adhere to zoning regulations.

b. Safety Assurance:

Obtaining permits involves inspections by qualified professionals. This oversight ensures that your electrical work is safe, reducing the risk of fire hazards, electric shocks, and other potential dangers.

2. Identify the Type of Permit Needed

a. Electrical Permit:

Specifically for electrical work, an electrical permit is necessary for projects like wiring upgrades, panel installations, or any modification to the electrical system.

b. Building Permit:

Some projects may require both an electrical permit and a building permit. Building permits are broader and may be needed for structural changes, such as adding a new room or altering walls.

3. Contact the Local Building Department

a. Research Local Requirements:

Visit the local building department's website or contact them directly to understand the specific permit requirements for your area. Each jurisdiction may have its own regulations.

b. Inquire about Necessary Documentation:

Ask about the documentation needed to apply for the permit. This may include detailed project plans, specifications, and sometimes contractor licensing information.

4. Prepare Required Documentation

a. Project Plans:

Create detailed plans outlining the scope of your electrical project. Include a circuit map, wiring diagrams, and any other relevant information that demonstrates compliance with safety standards.

b. Specifications:

Provide specifications for the materials you intend to use. This ensures that your choices meet safety and quality standards.

5. Submit the Permit Application

a. Complete the Application Form:

Fill out the permit application form accurately and completely. Ensure that all required information is provided, including your contact details, project scope, and any contractor information if applicable.

b. Pay the Permit Fee:

Most permits come with a fee. Be prepared to pay this fee when submitting your application. The fee may vary depending on the scope and complexity of your project.

6. Schedule and Attend Inspections

a. Inspection Process:

After obtaining the permit, the local building department will schedule inspections at key points during your project. This could include inspections for rough wiring, final installation, and safety checks.

b. Cooperate with Inspectors:

Cooperate with inspectors during visits. Address any issues they identify promptly to ensure compliance with regulations.

7. Keep Records of Permits and Inspections

a. Organize Documentation:

Keep a file with all permit-related documents, including the application form, project plans, and inspection records. This file serves as proof of compliance and may be valuable for future reference.

b. Display Permits:

If required, display the issued permits in a visible location on the job site. This indicates to inspectors and authorities that your project is in compliance.

Recap: Obtaining necessary permits involves understanding their importance, identifying the type of permit needed, contacting the local building department, preparing required documentation, submitting the permit application, scheduling and attending inspections, and keeping records of permits and inspections.

Chapter 5

Running New Wires

Selecting the Right Wire for the Job

In the symphony of home electrical projects, running new wires is like composing the melody – each note must be chosen carefully to ensure harmony. This chapter focuses on the art of selecting the right wire for the job, guiding you through the considerations that will transform your wiring endeavors into a seamless and efficient masterpiece.

1. Understand Wire Types

a. Non-Metallic (NM) Cable (Romex): The All-Purpose Player

Non-metallic cables, commonly known as Romex, consist of insulated conductors bundled together in a flexible plastic sheath. This versatile option is suitable for various residential applications and is easy to work with.

b. Armored Cable (AC): Robust Protection

Armored cables feature a metal sheath for added protection, making them suitable for areas where wiring may be exposed to physical damage. This robust choice is often used in commercial and industrial settings.

c. Conduit Wiring: The Customizable Shield

Conduit wiring involves individual wires or cables enclosed in a protective tube. Conduit provides an extra layer of protection and is ideal for customizable and exposed installations.

2. Gauge Matters: Choose the Right Size

a. Understanding Wire Gauge:

Wire gauge refers to the thickness of the wire. Thicker wires (lower gauge) can carry more current over longer distances without overheating. Understanding the electrical load helps determine the appropriate wire gauge for your project.

b. Ampacity Considerations:

Each wire gauge has a specific ampacity, indicating its safe current-carrying capacity. Ensure that the wire's ampacity matches or exceeds the current requirements of your devices to prevent overheating.

3. Insulation Material: Safeguarding Conductors

a. PVC (Polyvinyl Chloride): Common and Affordable
PVC is a widely used insulation material due to its affordability and versatility. It provides good protection against moisture and physical damage.

b. THHN/THWN: Heat and Moisture Resistance
Thermoplastic High Heat-Resistant Nylon (THHN) and Thermoplastic Heat and Water-Resistant Nylon (THWN) are insulation materials designed for enhanced heat and moisture resistance. They are suitable for conduit wiring.

4. Consider Environmental Factors

a. Indoor vs. Outdoor Wiring:
Outdoor wiring requires materials that can withstand exposure to the elements. UV-resistant jackets and waterproof insulation are essential for outdoor applications.

b. Wet vs. Dry Locations:
In wet locations, such as bathrooms or kitchens, use wiring materials with moisture-resistant features to prevent corrosion and reduce the risk of electrical issues.

5. Plan for Future Expansion

a. Consider Future Needs:
Anticipate any potential future additions or modifications to your electrical system. Choosing wiring that accommodates future expansion minimizes the need for rewiring.

b. Install Conduits for Flexibility:
Conduits offer flexibility and make future modifications easier. Installing conduits during the initial wiring process allows for the addition of new wires without significant disruption.

6. Code Compliance: Adherence to Standards

a. Familiarize Yourself with Local Codes:
Different regions have specific electrical codes. Familiarize yourself with local codes and regulations to ensure your wiring complies with safety and legal standards.

b. Consult with Professionals:
If in doubt, consult with a professional electrician. They can provide guidance on code compliance and help you make informed decisions.

Recap: Selecting the right wire involves understanding wire types, choosing the correct gauge, considering insulation materials, factoring in environmental conditions, planning for future expansion, and ensuring compliance with local electrical codes.

By selecting the right wire for each aspect of your project, you're orchestrating a composition of safety, efficiency, and flexibility. Your wiring choices will not only power your present needs but also harmonize with the potential of future expansions. Let's wire up your space with the precision and care that only the right selection of wires can provide.

Techniques for Fishing Wires Through Walls

In the grand tapestry of home electrical projects, fishing wires through walls is a skill that turns the invisible into the visible, seamlessly connecting the power within. This chapter dives into the art of wire fishing, revealing techniques that will empower you to navigate walls and ceilings with finesse, ensuring a smooth and efficient wiring process.

1. Use Existing Pathways

a. Electrical Boxes:

Utilize existing electrical boxes as entry points. Remove the cover plate and carefully thread the wire through the box, guiding it to the desired destination.

b. Cable Pathways:

Leverage pathways such as cable conduits, if available. These pre-existing channels facilitate wire fishing and reduce the need for invasive measures.

2. Employ Fish Tapes or Fish Rods

a. Fish Tapes:

Fish tapes are flexible, flat metal tapes designed for threading wires through walls. Attach the wire to the end of the tape and guide it through the wall by pushing or pulling the tape.

b. Fish Rods:

Fish rods, or fish sticks, are rigid rods that can be connected to create a longer reach. Attach the wire to the end and push or pull the rod through the wall to guide the wire.

3. Magnetic Wire Pulling Systems

a. Magnetic Fish Tape:

Magnetic fish tapes have a magnetized tip, allowing them to attract and guide the wire through the wall. This is particularly useful when dealing with drywall or other non-metallic materials.

b. Magnepull System:

The Magnepull system employs a magnetic connection between the pulling tool and a retriever, making it easier to navigate walls and pull wires with precision.

4. Vacuum Assisted Wire Pulling

a. Attach a String to a Vacuum Cleaner:

Tie a string to the end of the wire and use a vacuum cleaner to create suction. Guide the vacuum nozzle near the destination point, and the wire, attached to the string, will follow the vacuum's path.

b. Use a Blow-in Insulation Machine:

In some cases, where there's blow-in insulation, you can use a blowing machine to propel the wire through the wall. Ensure proper precautions are taken to avoid damage to the insulation.

5. Wall-Cavity Inspection and Planning

a. Identify Obstacles:

Before fishing wires, inspect the wall cavity to identify potential obstacles such as cross-bracing, fire blocks, or other obstructions. Knowing the layout helps plan the wire path.

b. Plan the Route:

Plan the route before starting. Identify the entry and exit points, and consider the shortest and most efficient path to minimize the effort required.

6. Techniques for Different Wall Types

a. Drywall:

When dealing with drywall, cut small openings for access. A stud finder can help locate wall studs, providing guidance for fishing wires between them.

b. Plaster Walls:

Plaster walls may require more care. Use a stud finder to locate wooden laths, and fish wires through the gaps between them.

7. Protective Measures

a. Protect the Wire:

To prevent damage to the wire, use protective measures such as conduit or PVC tubing in areas where the wire is vulnerable.

b. Be Mindful of Fire Blocking:

In some construction, fire blocking may be present between studs. Use techniques like drilling small holes to navigate wires through these barriers.

Recap: Techniques for fishing wires through walls involve using existing pathways, employing fish tapes or fish rods, utilizing magnetic wire pulling systems, exploring vacuum-assisted methods, conducting wall-cavity inspections, planning the route, adapting techniques for different wall types, and implementing protective measures.

Mastering wire fishing techniques is like unraveling a mystery within your walls. By employing these methods with precision and care, you'll seamlessly thread wires through the hidden spaces, bringing power and connectivity wherever it's needed. Let's navigate the walls with confidence and turn the invisible into the visible in your electrical projects.

Wiring for Different Fixtures and Appliances

In the electrifying journey of home improvement, wiring for various fixtures and appliances is the backbone that brings functionality and convenience to your space. This chapter delves into the specifics of wiring for different fixtures and appliances, guiding you through the intricacies of providing power to the elements that illuminate, control, and enhance your home.

1. Wiring for Light Fixtures

a. Ceiling Fixtures:

For ceiling fixtures, typically you'll run wiring from the electrical box in the ceiling to the fixture. Ensure compatibility between the fixture and the wattage rating of the wire.

b. Wall-Mounted Fixtures:

Wall-mounted fixtures require running wires vertically or horizontally within the wall. Ensure the wire is appropriately secured and protected to avoid damage.

2. Wiring for Outlets

a. Standard Outlets:

Wiring for standard outlets involves connecting the hot, neutral, and ground wires to the respective terminals on the outlet. Follow local codes and ensure proper grounding for safety.

b. GFCI Outlets:

Ground Fault Circuit Interrupter (GFCI) outlets provide additional safety by interrupting the circuit if a ground fault is detected. Wire GFCI outlets according to manufacturer instructions.

3. Wiring for Switches

a. Single-Pole Switches:

Single-pole switches control one light or group of lights from a single location. Wire these switches by connecting the hot wire to the common terminal and the switched hot wire to the other terminal.

b. Three-Way Switches:

Three-way switches control a light from two locations. Wiring involves a common terminal, two traveler terminals, and connecting the switched hot wire accordingly.

4. Wiring for Ceiling Fans

a. Mounting Box Preparation:
Ensure the ceiling fan is mounted on a properly rated electrical box. Follow manufacturer guidelines for box type and installation.

b. Connecting Wires:
Connect the fan's hot and neutral wires to the corresponding wires in the ceiling. If the fan has a light kit, wire it similarly to a ceiling light fixture.

5. Wiring for Appliances

a. Kitchen Appliances:
Kitchen appliances, such as ovens and dishwashers, often require dedicated circuits. Follow manufacturer instructions and ensure the wiring meets the appliance's power requirements.

b. Laundry Appliances:
Wiring for laundry appliances like washing machines and dryers involves dedicated circuits. Pay attention to voltage, amperage, and receptacle types specified by the appliances.

6. Wiring for Outdoor Fixtures

a. Weatherproof Outlets:
Outdoor outlets need to be weatherproof and equipped with appropriate covers. Use outdoor-rated wires and ensure proper grounding.

b. Landscape Lighting:
Wiring for landscape lighting may involve burying wires underground. Use outdoor-rated wires and protect connections with waterproof enclosures.

7. Wiring for HVAC Systems

a. Thermostats:
Wiring for thermostats involves connecting wires to designated terminals on the thermostat. Follow the thermostat manufacturer's wiring diagram.

b. Air Conditioners and Furnaces:
HVAC systems often require dedicated circuits. Follow manufacturer instructions for wiring and adhere to local codes for safety.

8. Wiring for Smart Home Devices

a. Smart Switches and Outlets:

Wiring for smart switches and outlets is similar to standard devices. Additionally, connect the device to a compatible smart home hub or network.

b. Smart Thermostats:

Smart thermostats may require a common wire (C-wire) for power. Follow the manufacturer's instructions for wiring and compatibility.

Recap: Wiring for different fixtures and appliances involves understanding the specific requirements of each element. Ensure proper grounding, follow manufacturer instructions, and adhere to local electrical codes for safety and functionality.

As you embark on wiring for various fixtures and appliances, think of it as orchestrating a symphony of connectivity. By following the correct wiring procedures for each element, you're not just providing power – you're creating a harmonious and functional environment within your home. Let's wire up your space with precision and ensure every element dances to the electrical melody.

Chapter 6

Installing Fixtures

Lighting Fixture Installation

In the radiant realm of home improvement, the installation of lighting fixtures is the crowning touch that illuminates and transforms spaces. This chapter is dedicated to the art of lighting fixture installation, guiding you through the steps to bring light to your home with precision and style.

1. Gather Necessary Tools and Materials

a. Essential Tools:

Before you begin, ensure you have the necessary tools, including a screwdriver, wire stripper, pliers, voltage tester, and a ladder if the fixture is ceiling-mounted.

b. Materials:

Gather the lighting fixture, wire nuts, mounting hardware, and any additional components specified by the manufacturer.

2. Turn Off Power

a. Safety First:

Ensure safety by turning off the power to the circuit at the electrical panel. Confirm the power is off using a voltage tester before proceeding.

3. Remove Old Fixture (If Applicable)

a. Disconnect Wires:

If replacing an existing fixture, disconnect the wires from the old fixture. Note the wire connections or take a picture for reference.

b. Remove Mounting Hardware:

Unscrew and remove the mounting hardware that secures the old fixture to the electrical box.

4. Mounting Bracket Installation

a. Secure Mounting Bracket:

If your new fixture requires a mounting bracket, install it according to the manufacturer's instructions. This bracket provides a stable base for the fixture.

5. Connect Wires

a. Match Wires:

Match the wires from the fixture to the corresponding wires in the electrical box. Typically, this involves connecting the black (hot) wire to the black wire, white (neutral) to white, and ground to ground.

b. Use Wire Nuts:

Secure the wire connections using wire nuts. Twist the wires together, cover with a wire nut, and wrap with electrical tape for added security.

6. Attach the Fixture

a. Secure the Fixture:

Align the fixture with the mounting bracket or electrical box and secure it in place using the provided screws or nuts.

b. Follow Manufacturer's Instructions:

Adhere to the manufacturer's instructions regarding the specific attachment method for your fixture. Some fixtures may have different mounting mechanisms.

7. Install Bulbs and Shades

a. Install Bulbs:

Insert the appropriate light bulbs into the fixture, following the wattage recommendations specified by the manufacturer.

b. Attach Shades or Covers:

If your fixture includes shades or covers, attach them securely according to the provided instructions.

8. Double-Check Connections

a. Verify Tight Connections:

Double-check that all wire connections are tight and secure. This ensures a reliable electrical connection and prevents potential issues.

9. Restore Power and Test

a. Restore Power:

Turn the power back on at the electrical panel. If the fixture has a built-in switch, ensure it is in the "off" position.

b. Test the Fixture:

Test the lighting fixture by flipping the switch. Verify that the fixture functions as expected and that the light illuminates properly.

10. Final Adjustments and Cleanup

a. Adjust Positioning:

Make any final adjustments to the positioning of the fixture to ensure it is level and aligned correctly.

b. Cleanup:

Dispose of old fixtures responsibly and clean up any debris from the installation process.

Recap: Lighting fixture installation involves gathering necessary tools and materials, turning off power, removing old fixtures if applicable, installing a mounting bracket, connecting wires, attaching the fixture securely, installing bulbs and shades, double-checking connections, restoring power and testing, and making final adjustments.

As you complete the installation of your lighting fixtures, envision the transformation of your space with each flicker of light. By following these steps with care and precision, you're not just installing fixtures – you're bringing warmth, ambiance, and functionality to your home. Let the light shine brightly on your well-executed lighting fixture installations.

Ceiling Fan Installation

In the realm of home comfort and aesthetics, installing a ceiling fan is a dual-purpose endeavor, providing both functionality and a stylish touch. This chapter delves into the intricacies of ceiling fan installation, guiding you through the step-by-step process to ensure a breeze of comfort and elegance in your living spaces.

1. Gather Necessary Tools and Materials

a. Essential Tools:

Before you begin, gather tools such as a screwdriver, wire stripper, pliers, adjustable wrench, and a ladder. Ensure you have the appropriate tools for the specific fan model.

b. Materials:

Collect the ceiling fan kit, including the fan blades, mounting bracket, screws, wire nuts, and any additional components mentioned in the manufacturer's instructions.

2. Turn Off Power

a. Safety First:

Ensure safety by turning off the power to the ceiling fan circuit at the electrical panel. Verify the power is off using a voltage tester before proceeding.

3. Assemble the Ceiling Fan

a. Follow Manufacturer's Instructions:
Consult the manufacturer's instructions for assembling the ceiling fan. This usually involves attaching the fan blades to the motor housing and assembling the downrod.

b. Secure the Downrod:
Attach the downrod to the motor housing and secure it with the provided screws. The length of the downrod depends on the height of your ceiling.

4. Mounting Bracket Installation

a. Locate Ceiling Joist:
Use a stud finder to locate the ceiling joist where you'll mount the fan. The mounting bracket must be secured to a joist for stability.

b. Secure Mounting Bracket:
Attach the mounting bracket to the ceiling using the provided screws. Ensure it is securely fastened to the ceiling joist.

5. Wiring Connections

a. Expose Ceiling Wires:
Remove the canopy cover at the top of the fan to expose the ceiling wires. The canopy cover is typically held in place by screws.

b. Connect Wires:
Connect the wires from the fan (usually colored black, white, and green or bare) to the corresponding wires in the ceiling. Secure connections with wire nuts and cover with the provided canopy.

6. Hang the Ceiling Fan

a. Lift the Fan:
With the help of an assistant, lift the assembled fan and connect it to the mounting bracket. Align the holes in the fan's mounting bracket with the screws on the ceiling bracket.

b. Secure the Fan:
Secure the fan to the ceiling bracket using the provided screws. Ensure the fan is balanced and tightly secured.

7. Attach Fan Blades

a. Follow Blade Installation Instructions:

Follow the manufacturer's instructions for attaching the fan blades. Typically, this involves aligning the blades with the pre-drilled holes on the motor housing and securing them with screws.

8. Install Light Kit (If Applicable)

a. Attach Light Kit:
If your ceiling fan comes with a light kit, follow the manufacturer's instructions to attach it. Connect the wiring according to the provided diagram.

b. Install Bulbs:
Insert the appropriate light bulbs into the light kit, following the wattage recommendations.

9. Secure Fan Pull Chains or Remote Control

a. Attach Pull Chains:
If your fan has pull chains, attach them according to the instructions. Ensure they are accessible for easy fan and light control.

b. Install Remote Control (If Applicable):
If your fan comes with a remote control, follow the instructions to install and pair it with the fan.

10. Test the Ceiling Fan

a. Restore Power:

Turn the power back on at the electrical panel. Use the fan's pull chains or remote control to test the fan's operation and light functionality.

b. Check for Wobbling:

If the fan wobbles, use a balancing kit (if provided) to correct the balance. Ensure all screws are tightened securely.

11. Final Adjustments and Cleanup

a. Adjust Settings:

Adjust the fan settings, such as speed and direction, according to your preference.

b. Cleanup:

Dispose of packaging materials and clean up the installation area. Enjoy your newly installed ceiling fan!

Recap: Ceiling fan installation involves gathering necessary tools and materials, turning off power, assembling the fan, installing the mounting bracket, making wiring connections, hanging the fan, attaching fan blades, installing a light kit if applicable, securing fan pull chains or a remote control, testing the fan, and making final adjustments.

As you complete the installation of your ceiling fan, imagine the cool breeze and added charm it brings to your space. By following these steps diligently, you're not just installing a fan, you're enhancing the comfort and ambiance of your home. Let the gentle whirl of the ceiling fan be a testament to your well-executed installation.

Wiring for Other Electrical Fixtures

Beyond lighting and ceiling fans, your home may host a variety of other electrical fixtures that contribute to its functionality and aesthetics. This chapter explores the wiring considerations for various fixtures such as outlets, switches, and specialty electrical elements, guiding you through the process of ensuring a reliable and safe electrical connection.

1. Wiring for Electrical Outlets

a. Standard Outlets:

Standard outlets typically have three connections: hot (black), neutral (white), and ground (green or bare). Connect these wires to the corresponding terminals on the outlet.

b. GFCI Outlets:

Ground Fault Circuit Interrupter (GFCI) outlets add an extra layer of safety. Wire GFCI outlets according to the manufacturer's instructions, ensuring proper line and load connections.

2. Wiring for Switches

a. Single-Pole Switches:
Single-pole switches control one light or group of lights from a single location. Wire these switches by connecting the hot wire to the common terminal and the switched hot wire to the other terminal.

b. Three-Way Switches:
Three-way switches control a light from two locations. Wiring involves a common terminal, two traveler terminals, and connecting the switched hot wire accordingly.

3. Wiring for Specialty Fixtures

a. Dimmer Switches:
Dimmer switches allow control over the brightness of lights. Follow the manufacturer's instructions to wire dimmer switches, usually connecting hot, neutral, and ground wires.

b. Timer Switches:
Timer switches automate the turning on and off of lights. Wire timer switches according to the provided instructions, connecting hot, neutral, and ground wires.

4. Wiring for Specialty Outlets

a. USB Outlets:

USB outlets include ports for charging devices. Wire USB outlets by connecting hot, neutral, and ground wires as specified by the manufacturer.

b. Smart Outlets:

Smart outlets connect to a home automation system. Follow the manufacturer's instructions for wiring and connecting the outlet to the smart home network.

5. Wiring for Exhaust Fans

a. Mounting Box Preparation:

Ensure the exhaust fan is mounted on a properly rated electrical box. Follow manufacturer guidelines for box type and installation.

b. Connect Wires:

Connect the fan's hot, neutral, and ground wires to the corresponding wires in the ceiling. Follow the manufacturer's instructions for proper wiring.

6. Wiring for Doorbell Systems

a. Transformer Connection:
Doorbell systems often include a transformer. Connect the wires from the doorbell button to the transformer and from the transformer to the chime unit.

b. Chime Unit Wiring:
Connect the chime unit to the transformer and doorbell button wires, following the manufacturer's instructions.

7. Wiring for Thermostats

a. Identify Wires:
Identify the wires connected to your existing thermostat. Common wires include R (power), W (heat), Y (cooling), G (fan), and C (common).

b. Connect to New Thermostat:
Connect the identified wires to the corresponding terminals on the new thermostat, following the manufacturer's wiring diagram.

8. Wiring for Security Cameras

a. Power and Video Wires:
Security cameras typically have power and video wires. Connect the power wires to a suitable power source, and

connect the video wires to a recording or monitoring device.

b. Follow Manufacturer Instructions:

Follow the manufacturer's instructions for wiring and configuring the security camera system.

9. Wiring for Specialty Fixtures (e.g., Sconces, Under-Cabinet Lighting)

a. Follow Fixture Instructions:

Each specialty fixture may have unique wiring requirements. Follow the manufacturer's instructions for connecting hot, neutral, and ground wires.

b. Install Junction Boxes if Needed:

If additional wiring connections are required, install junction boxes and connect the wires according to local electrical codes.

Recap: Wiring for other electrical fixtures involves understanding the specific requirements of each fixture type. Ensure proper grounding, follow manufacturer instructions, and adhere to local electrical codes for safety and functionality.

As you navigate the wiring for various electrical fixtures, consider it a personalized orchestration of functionality in your home. By adhering to proper wiring practices and following manufacturer instructions, you're not just connecting wires – you're creating a seamless and reliable electrical network that enhances the usability and convenience of your living space. Let the electrical fixtures in your home operate harmoniously under your well-executed wiring.

Chapter 7

Replacing Old Outlets and Switches

Identifying and Assessing Old Wiring

Beyond lighting and ceiling fans, your home may host a variety of other electrical fixtures that contribute to its functionality and aesthetics. This chapter explores the wiring considerations for various fixtures such as outlets, switches, and specialty electrical elements, guiding you through the process of ensuring a reliable and safe electrical connection.

1. Wiring for Electrical Outlets

a. Standard Outlets:
   Standard outlets typically have three connections: hot (black), neutral (white), and ground (green or bare). Connect these wires to the corresponding terminals on the outlet.

b. GFCI Outlets:

Ground Fault Circuit Interrupter (GFCI) outlets add an extra layer of safety. Wire GFCI outlets according to the manufacturer's instructions, ensuring proper line and load connections.

2. Wiring for Switches

a. Single-Pole Switches:

Single-pole switches control one light or group of lights from a single location. Wire these switches by connecting the hot wire to the common terminal and the switched hot wire to the other terminal.

b. Three-Way Switches:

Three-way switches control a light from two locations. Wiring involves a common terminal, two traveler terminals, and connecting the switched hot wire accordingly.

3. Wiring for Specialty Fixtures

a. Dimmer Switches:

Dimmer switches allow control over the brightness of lights. Follow the manufacturer's instructions to wire dimmer switches, usually connecting hot, neutral, and ground wires.

b. Timer Switches:

Timer switches automate the turning on and off of lights. Wire timer switches according to the provided instructions, connecting hot, neutral, and ground wires.

4. Wiring for Specialty Outlets

a. USB Outlets:

USB outlets include ports for charging devices. Wire USB outlets by connecting hot, neutral, and ground wires as specified by the manufacturer.

b. Smart Outlets:

Smart outlets connect to a home automation system. Follow the manufacturer's instructions for wiring and connecting the outlet to the smart home network.

5. Wiring for Exhaust Fans

a. Mounting Box Preparation:

Ensure the exhaust fan is mounted on a properly rated electrical box. Follow manufacturer guidelines for box type and installation.

b. Connect Wires:

Connect the fan's hot, neutral, and ground wires to the corresponding wires in the ceiling. Follow the manufacturer's instructions for proper wiring.

6. Wiring for Doorbell Systems

a. Transformer Connection:
Doorbell systems often include a transformer. Connect the wires from the doorbell button to the transformer and from the transformer to the chime unit.

b. Chime Unit Wiring:
Connect the chime unit to the transformer and doorbell button wires, following the manufacturer's instructions.

7. Wiring for Thermostats

a. Identify Wires:
Identify the wires connected to your existing thermostat. Common wires include R (power), W (heat), Y (cooling), G (fan), and C (common).

b. Connect to New Thermostat:
Connect the identified wires to the corresponding terminals on the new thermostat, following the manufacturer's wiring diagram.

8. Wiring for Security Cameras

a. Power and Video Wires:
Security cameras typically have power and video wires. Connect the power wires to a suitable power source, and

connect the video wires to a recording or monitoring device.

b. Follow Manufacturer Instructions:
Follow the manufacturer's instructions for wiring and configuring the security camera system.

9. Wiring for Specialty Fixtures (e.g., Sconces, Under-Cabinet Lighting)

a. Follow Fixture Instructions:
Each specialty fixture may have unique wiring requirements. Follow the manufacturer's instructions for connecting hot, neutral, and ground wires.

b. Install Junction Boxes if Needed:
If additional wiring connections are required, install junction boxes and connect the wires according to local electrical codes.

Recap: Wiring for other electrical fixtures involves understanding the specific requirements of each fixture type. Ensure proper grounding, follow manufacturer instructions, and adhere to local electrical codes for safety and functionality.

As you navigate the wiring for various electrical fixtures, consider it a personalized orchestration of functionality in your home. By adhering to proper wiring practices and following manufacturer instructions, you're not just connecting wires, you're creating a seamless and reliable electrical network that enhances the usability and convenience of your living space. Let the electrical fixtures in your home operate harmoniously under your well-executed wiring.

Upgrading Outlets to GFCI

In the pursuit of enhanced electrical safety, upgrading outlets to Ground Fault Circuit Interrupter (GFCI) receptacles is a prudent and crucial step. This chapter will guide you through the process of upgrading your outlets to GFCI, providing an extra layer of protection against electrical hazards.

1. Identify Outlets to Upgrade

a. Determine Appropriate Locations:
GFCI outlets are essential in areas prone to moisture, such as kitchens, bathrooms, outdoor spaces, garages, and basements. Identify the outlets in these locations that need upgrading.

2. Turn Off Power

a. Safety First:
Turn off the power to the circuit at the electrical panel that supplies electricity to the outlets you're upgrading. Confirm the power is off using a voltage tester.

3. Remove the Old Outlet

a. Unscrew and Disconnect:

Unscrew the outlet cover plate and the screws securing the outlet to the electrical box. Carefully pull the outlet out, ensuring the wires are still attached.

b. Take Note of Wiring:

Note the arrangement of wires on the old outlet. Typically, there are hot (black), neutral (white), and ground (green or bare) wires.

4. Install the GFCI Outlet

a. Match Wires:

Match the wires from the electrical box to the corresponding terminals on the GFCI outlet. Connect the hot wire to the brass or gold screw, the neutral wire to the silver screw, and the ground wire to the green screw.

b. Attach Wires Securely:

Securely attach the wires to the GFCI outlet by tightening the terminal screws. Ensure a snug fit to prevent loose connections.

5. Mount the GFCI Outlet

a. Position in the Electrical Box:

Carefully position the GFCI outlet back into the electrical box, ensuring the wires are neatly arranged and do not interfere with the outlet's functionality.

b. Screw in Place:

Secure the GFCI outlet to the electrical box by screwing it in using the screws provided with the outlet. Make sure it is flush with the wall.

6. Test the GFCI Outlet

a. Restore Power:

Turn the power back on at the electrical panel. The GFCI outlet should have a built-in indicator light that turns on to indicate power.

b. Test Button:

Press the "Test" button on the GFCI outlet. The indicator light should turn off, signifying that the GFCI is working correctly.

c. Reset Button:

Press the "Reset" button to restore power to the GFCI outlet. The indicator light should turn back on.

7. Test Additional Outlets in the Circuit

a. Verify Protection:

Plug a lamp or small appliance into the GFCI outlet and test its functionality. Additionally, test other outlets downstream from the GFCI to ensure they are protected.

8. Install GFCI Outlet Covers (If Needed)

a. Extra Weather Protection:

In outdoor locations, consider installing weather-resistant GFCI outlet covers to provide extra protection against the elements.

9. Label the GFCI Outlet

a. Clearly Indicate GFCI Protection:

Adhere to National Electrical Code (NEC) requirements by labeling the GFCI outlet to indicate that it provides downstream protection. Use the provided "GFCI Protected" stickers.

Recap: Upgrading outlets to GFCI involves identifying appropriate locations, turning off power, removing the old outlet, installing the GFCI outlet, mounting it securely, testing its functionality, checking additional outlets in the circuit, installing covers if needed, and labeling the GFCI outlet for clear identification.

Upgrading outlets to GFCI is a proactive measure in ensuring the safety of your home's electrical system, especially in areas with increased risk. By following these steps meticulously, you're not just upgrading outlets – you're enhancing the safety net of your electrical infrastructure. Let the added protection of

GFCI outlets be a testament to your commitment to electrical safety.

Upgrading Switches for Safety

In the pursuit of an electrical environment that prioritizes safety and convenience, upgrading switches is a significant step. This chapter will guide you through the process of upgrading switches, ensuring not only a reliable operation but also an enhanced level of safety within your living spaces.

1. Identify Switches to Upgrade

a. Assess High-Traffic Areas:

Identify switches in high-traffic areas, such as hallways, stairwells, and frequently used rooms, for upgrading. Consider switches that show signs of wear, damage, or inconsistent performance.

2. Turn Off Power

a. Safety First:

Turn off the power to the circuit at the electrical panel that supplies electricity to the switches you're upgrading. Confirm the power is off using a voltage tester.

3. Remove the Old Switch

a. Unscrew and Disconnect:

Unscrew the switch cover plate and the screws securing the switch to the electrical box. Carefully pull the switch out, ensuring the wires are still attached.

b. Take Note of Wiring:

Note the arrangement of wires on the old switch. Typically, switches have hot (black), traveler (for three-way switches), and ground (green or bare) wires.

4. Install the Upgraded Switch

a. Match Wires:

Match the wires from the electrical box to the corresponding terminals on the upgraded switch. Connect the hot wire to the common terminal, travelers to their respective terminals, and the ground wire to the green screw.

b. Attach Wires Securely:

Securely attach the wires to the upgraded switch by tightening the terminal screws. Ensure a snug fit to prevent loose connections.

5. Mount the Upgraded Switch

a. Position in the Electrical Box:

Carefully position the upgraded switch back into the electrical box, ensuring the wires are neatly arranged and do not interfere with the switch's functionality.

b. Screw in Place:

Secure the upgraded switch to the electrical box by screwing it in using the screws provided with the switch. Make sure it is flush with the wall.

6. Test the Upgraded Switch

a. Restore Power:

Turn the power back on at the electrical panel.

b. Test Switch Operation:

Test the functionality of the upgraded switch by toggling it on and off. Ensure smooth and consistent operation.

7. Assess Additional Safety Features

a. Consider Motion Sensors:

In high-traffic areas, consider upgrading switches to include motion sensors. These switches automatically turn lights on or off based on motion detection, enhancing both safety and energy efficiency.

b. Explore Timers or Smart Switches:
For enhanced control and safety, explore switches with timer features or smart switches that allow remote control and automation.

8. Label the Upgraded Switch (If Needed)

a. Clearly Indicate New Features:
If the upgraded switch has additional features or functions, consider labeling it for clear identification, especially if it includes unique capabilities like timers or motion sensors.

9. Regular Maintenance and Inspection

a. Periodic Checks:
Implement a routine schedule for inspecting switches and their performance. Look out for any signs of wear, damage, or inconsistencies.

Recap: Upgrading switches for safety involves identifying switches to upgrade, turning off power, removing the old switch, installing the upgraded switch, mounting it securely, testing its functionality, assessing additional safety features, labeling the switch if needed, and incorporating regular maintenance and inspection.

Upgrading switches is not just about modernizing your home; it's a commitment to a safer and more convenient living environment. By following these steps diligently, you're not merely upgrading switches – you're elevating the safety standards within your home. Let the upgraded switches be a testament to your dedication to a secure and efficient electrical infrastructure.

Chapter 8

Ensuring Safety to Code

Compliance with Electrical Codes

In the labyrinth of home electrical projects, adherence to electrical codes is the compass that ensures a journey marked by safety, reliability, and compliance. This chapter navigates the essential terrain of electrical codes, guiding you through the process of ensuring your electrical work aligns with established standards for a secure and code-compliant home.

1. Understanding Electrical Codes

a. Purpose and Importance:
Electrical codes are a set of regulations and standards established by regulatory bodies to ensure the safe installation, operation, and maintenance of electrical systems. Understanding the purpose and importance of these codes sets the foundation for a secure electrical environment.

b. Local Variations:

Recognize that electrical codes may vary by region and jurisdiction. Familiarize yourself with the specific codes applicable to your location, which are often based on the National Electrical Code (NEC) in the United States.

2. Staying Informed and Updated

a. Code Changes:

Electrical codes evolve to address emerging technologies and safety considerations. Stay informed about updates and revisions to the electrical codes relevant to your area. Check with local authorities or online resources for the latest code versions.

b. Professional Guidance:

Seek guidance from qualified professionals, such as electricians or inspectors, to understand the nuances of code compliance. Professionals are well-versed in local codes and can provide valuable insights.

3. Plan Review and Inspections

a. Pre-Project Consultation:

Before undertaking significant electrical projects, consider consulting with local authorities or inspectors. They can provide insights into code requirements specific to your project.

b. Inspection Requests:

Schedule inspections at key milestones during your project. This ensures that each phase complies with the relevant codes. Be prepared to address any issues raised during inspections.

4. Key Code Considerations

a. Wiring Methods:

Understand the approved wiring methods outlined in electrical codes. This includes guidelines for running wires, securing cables, and using conduits to protect electrical conductors.

b. Outlet and Switch Placement:

Electrical codes specify the appropriate placement of outlets and switches to ensure accessibility and convenience. Adhere to guidelines regarding spacing and height requirements.

5. Grounding and Bonding

a. Grounding Systems:

Electrical codes mandate proper grounding to safeguard against electrical faults. Understand the requirements for grounding systems, including grounding electrodes and conductors.

b. Bonding Metal Components:

Bonding metal components, such as plumbing and structural elements, is crucial for safety. Adhere to electrical codes that define the bonding requirements to prevent potential hazards.

6. Load Calculations and Circuits

a. Electrical Load:

Codes provide guidelines for calculating electrical loads to prevent overloading circuits. Understand how to perform load calculations and ensure circuits are appropriately sized for the connected devices.

b. Dedicated Circuits:

Identify devices that require dedicated circuits, such as kitchen appliances and HVAC systems. Comply with code requirements to prevent circuit overload and ensure safety.

7. Hazardous Locations and Equipment

a. Identify Hazardous Areas:

Electrical codes define hazardous locations, such as those prone to moisture or combustible dust. Implement code-compliant solutions, such as using moisture-resistant materials or explosion-proof equipment in designated areas.

b. Outdoor Installations:

Outdoor electrical installations must adhere to specific codes. Use weatherproof materials, outdoor-rated outlets, and proper enclosures to protect electrical components from the elements.

8. Documentation and Record-Keeping

a. Record Project Details:

Maintain detailed records of your electrical projects, including plans, permits, inspection reports, and a list of materials used. Proper documentation is valuable for future reference and potential property transactions.

b. Code Compliance Certificates:

Upon project completion, obtain a code compliance certificate if applicable in your area. This formal document attests that the electrical work meets the requirements of the relevant codes.

Recap: Ensuring safety to code involves understanding electrical codes, staying informed and updated, undergoing plan reviews and inspections, considering key code aspects, addressing grounding and bonding requirements, performing load calculations, identifying hazardous locations, and maintaining thorough documentation.

In the realm of electrical work, compliance with electrical codes is not just a legal requirement; it's a commitment to the safety and well-being of your household. By navigating the intricate landscape of codes with diligence and awareness, you're not merely meeting standards, you're elevating the electrical integrity of your home to the highest level of safety and reliability. Let code compliance be the cornerstone of your electrical endeavors.

Grounding and Bonding Requirements

Proper grounding and bonding are fundamental pillars of electrical safety, creating a reliable path for electricity to follow in the event of a fault. This section delves into the essential grounding and bonding requirements, providing a roadmap for a secure and code-compliant electrical system in your home.

1. Understanding Grounding:

a. Purpose:

The primary purpose of grounding is to provide a safe pathway for electric currents to flow into the ground in the event of a fault. This helps prevent electrical shock, protects appliances, and ensures the safe operation of electrical systems.

b. Components:

Grounding involves connecting electrical systems to the earth through grounding electrodes, such as ground rods or metal water pipes. Conductors, typically made of copper or aluminum, facilitate this connection.

2. Bonding Metal Components:

a. Definition:
Bonding is the practice of connecting all metal components of an electrical system to ensure they are at the same electrical potential. This prevents potential differences that could lead to electric shock or fire hazards.

b. Examples of Bonding:
Bonding includes connecting metal electrical boxes, conduit systems, equipment enclosures, and grounding conductors to create a unified and interconnected system.

3. Grounding Electrodes:

a. Types:
Common grounding electrodes include ground rods, metal water pipes, and concrete-encased electrodes. The selection depends on local codes and the nature of the electrical system.

b. Installation:
Proper installation involves driving ground rods into the earth, connecting them with grounding conductors, and ensuring low-resistance paths for fault currents to dissipate safely.

4. Grounding Conductors:

a. Size and Material:

Grounding conductors must have sufficient size and material to carry fault currents safely. Copper or aluminum conductors are commonly used, and their size is determined based on the electrical system's specifications.

b. Connection Points:

Grounding conductors connect to various components, including electrical panels, equipment enclosures, and grounding electrodes. These connections must be secure and low-resistance to ensure effective grounding.

5. Bonding of Service Equipment:

a. Main Service Panel:

The main service panel must have bonding between the neutral (grounded) conductor and the grounding conductor. This connection is typically done at the main disconnect.

b. Grounding Electrode System:

All grounding electrodes, including ground rods and metal water pipes, should be bonded together to create a continuous and low-resistance path to the earth.

6. Equipment Bonding:

a. Metal Equipment Enclosures:
Metal equipment enclosures, such as electrical boxes and panels, must be bonded to the grounding system. This ensures that any metal part remains at the same electrical potential.

b. Equipment Grounding Conductors:
Equipment grounding conductors, typically green or bare, connect metal parts of equipment to the grounding system, creating a bond that mitigates shock hazards.

7. Grounding in Subpanels:

a. Separate Ground and Neutral:
In subpanels, the grounding and neutral conductors must be kept separate. The neutral bus bar is bonded to the grounding system at the main service panel, but in subpanels, they remain isolated.

b. Grounding Electrode Conductor:
Subpanels also require a grounding electrode conductor to connect to local grounding electrodes, ensuring continuity with the main grounding system.

8. Grounding for Specific Installations:

a. Swimming Pools and Spas:

Specific grounding requirements exist for swimming pools and spas to prevent electrical hazards near water. Bonding metallic components, including pool shells and equipment, is crucial.

b. Lightning Protection Systems:

Lightning protection systems require grounding to safely dissipate the immense energy associated with lightning strikes. Grounding electrodes and conductors are strategically installed to create a low-resistance path.

9. Regular Maintenance and Inspection:

a. Visual Inspections:

Periodic visual inspections of grounding and bonding components are essential. Look for signs of corrosion, damage, or loose connections that could compromise the effectiveness of the system.

b. Testing:

Conducting resistance tests on grounding electrodes and conductors ensures their ability to carry fault currents. Regular testing helps identify any degradation or deterioration that may occur over time.

Recap: Grounding and bonding requirements are foundational for electrical safety. Understanding the purpose, components, and proper installation of grounding and bonding systems is crucial for creating a secure electrical environment. Regular maintenance and adherence to local codes contribute to the ongoing safety and reliability of the electrical system in your home.

Conducting Electrical Inspections

Electrical inspections are a crucial aspect of maintaining a safe and reliable electrical system in your home. This section provides a comprehensive guide on conducting electrical inspections, empowering you to identify potential hazards, ensure compliance with codes, and uphold the integrity of your electrical infrastructure.

1. Schedule Regular Inspections:

a. Frequency:

Establish a routine schedule for electrical inspections. Consider conducting annual inspections or more frequently if your home is older, or you've recently undertaken electrical projects.

b. Professional Inspections:

For a thorough evaluation, engage the services of a qualified electrician or inspector. Professionals bring expertise and can identify issues that may not be apparent to the untrained eye.

2. Visual Inspection Checklist:

a. Outlets and Switches:

Check all outlets and switches for signs of wear, damage, or discoloration. Ensure that outlets are secure and not loose.

b. Wiring and Cables:

Inspect wiring for any exposed or damaged sections. Look for frayed insulation, which could pose a fire hazard. Check for secure cable connections at outlets and switches.

c. Electrical Panels:

Examine the electrical panel for signs of overheating, burnt smells, or loose connections. Verify that circuit breakers or fuses are appropriately sized and labeled.

d. Lighting Fixtures:

Inspect lighting fixtures for loose bulbs, damaged wiring, or signs of overheating. Ensure that light switch plates are intact.

e. Appliances and Equipment:

Check the power cords of appliances and equipment for any damage. Verify that appliances are connected to grounded outlets.

3. Testing and Measurements:

a. Outlet Testing:

Use a receptacle tester to check the functionality of outlets. Ensure proper wiring with correct polarity, grounding, and GFCI protection where required.

b. Voltage Testing:

Conduct voltage tests to ensure a stable power supply. Irregular voltage levels can damage sensitive electronics and appliances.

c. Ground Fault Circuit Interrupters (GFCIs):

Test GFCI outlets to confirm they trip properly in the presence of a ground fault. Follow the manufacturer's instructions for testing procedures.

4. Inspection of Grounding and Bonding:

a. Grounding Electrodes:

Inspect grounding electrodes, such as ground rods, for signs of corrosion or damage. Ensure a low-resistance path to the earth.

b. Bonding of Metal Components:

Verify the bonding of metal components, including equipment enclosures and conduit systems. Ensure that all metal parts are at the same electrical potential.

5. Review of Circuit Load:

a. Load Calculation:

Assess the load on each circuit to ensure it does not exceed the circuit's capacity. Overloaded circuits can lead to overheating and pose fire risks.

b. Dedicated Circuits:

Confirm that high-demand appliances and equipment, such as refrigerators or air conditioners, are on dedicated circuits to prevent overloading.

6. Outdoor Electrical Inspection:

a. Weatherproofing:

Check outdoor outlets and lighting fixtures for proper weatherproofing. Ensure that covers are intact and provide protection against the elements.

b. Landscaping Interference:

Trim vegetation and clear debris around outdoor electrical components to prevent potential hazards and ensure accessibility.

7. Inspection of Specialized Systems:

a. Solar Panels:

If your home has solar panels, inspect the wiring and connections. Ensure that the system is properly grounded and complies with local regulations.

b. Smart Home Systems:

If you have a smart home system, check the wiring and connections. Ensure that devices and components are functioning as intended.

8. Documentation and Record-Keeping:

a. Maintain Records:

Keep a record of all inspections, including dates, findings, and any corrective actions taken. This documentation is valuable for future reference and potential property transactions.

b. Code Compliance Certificates:

If applicable, obtain code compliance certificates for electrical work. These certificates affirm that the work meets the requirements of relevant electrical codes.

9. Immediate Repairs and Corrections:

a. Address Hazards Promptly:
If any hazards or code violations are identified during the inspection, address them promptly. Immediate repairs are essential to prevent potential risks.

b. Professional Assistance:
For complex issues or if you are unsure about any findings, seek professional assistance. Electricians or inspectors can provide guidance and ensure that corrections are made effectively.

Recap: Conducting electrical inspections involves scheduling regular assessments, performing visual checks of outlets, switches, wiring, and panels, testing outlets and voltage, inspecting grounding and bonding, reviewing circuit loads, assessing outdoor electrical components, inspecting specialized systems, maintaining documentation, and addressing repairs promptly. These inspections contribute to the ongoing safety and reliability of your home's electrical system.

Chapter 9

Troubleshooting and Maintenance

Common Electrical Issues

Maintenance and troubleshooting are essential components of responsible home ownership, ensuring your electrical system operates efficiently and safely. This chapter provides insights into common electrical issues and regular maintenance practices, empowering you to identify, address, and prevent potential problems.

1. Tripped Circuit Breakers:

a. Cause:

Overloaded circuits or a short circuit can trip circuit breakers. Identify the cause by unplugging devices or reducing the load on the circuit.

b. Solution:

Reset the circuit breaker by moving it to the "off" position and then back to "on." If the issue persists, consult a professional to assess and rectify the problem.

2. Flickering Lights:

a. Cause:

Loose bulbs, faulty wiring, or issues with the electrical service can cause lights to flicker. Identify the source of the problem by checking connections and fixtures.

b. Solution:

Tighten loose bulbs, check for loose wiring, and ensure a stable power supply. If flickering persists, consult an electrician to investigate and address the underlying issue.

3. Dead Outlets:

a. Cause:

Dead outlets may result from tripped circuit breakers, faulty outlets, or wiring issues. Identify the cause by testing nearby outlets and checking for loose wiring.

b. Solution:

Reset tripped breakers, replace faulty outlets, or address wiring problems. If the issue persists, seek professional assistance to diagnose and resolve the problem.

4. High Energy Bills:

a. Cause:

Inefficient appliances, poor insulation, or outdated electrical systems can contribute to high energy bills. Evaluate energy usage patterns and the efficiency of appliances.

b. Solution:

Upgrade to energy-efficient appliances, improve insulation, and consider a home energy audit. Addressing inefficiencies can lead to significant energy savings.

5. Electric Shocks:

a. Cause:

Faulty wiring, damaged appliances, or poor grounding can lead to electric shocks. Identify the source by inspecting wiring and checking appliances.

b. Solution:

Address faulty wiring, repair or replace damaged appliances, and ensure proper grounding. If electric shocks persist, consult an electrician for a thorough inspection.

Regular Maintenance Practices

Proactive Measures for a Reliable Electrical System

1. Visual Inspections:

a. Regular Checks:
Periodically inspect outlets, switches, and electrical panels for signs of wear, damage, or discoloration. Identify issues early to prevent potential hazards.

b. Professional Inspections:
Engage a professional electrician for comprehensive annual inspections. Professionals can identify hidden issues and ensure compliance with electrical codes.

2. Testing and Measurements:

a. Outlet Testing:
Regularly use a receptacle tester to ensure outlets are wired correctly and provide proper grounding. Replace faulty outlets promptly.

b. Voltage Testing:
Conduct voltage tests to identify irregularities. Consistent voltage levels contribute to the longevity of appliances and prevent damage.

3. Tightening Connections:

a. Secure Fixtures:
Regularly check and tighten loose bulbs, outlet covers, and switch plates. Secure connections prevent flickering lights and potential hazards.

b. Panel Inspection:
Inspect the electrical panel for loose connections. Tighten screws and ensure secure connections to prevent overheating.

4. Cleaning and Dusting:

a. Dust Removal:
Keep electrical components, such as panels and outlets, free of dust. Dust accumulation can lead to overheating and reduce the efficiency of electrical systems.

b. Ventilation:
Ensure proper ventilation around electrical components. Adequate airflow helps dissipate heat and prevents components from overheating.

5. Surge Protection:

a. Install Surge Protectors:
  Safeguard sensitive electronics by installing surge protectors. These devices prevent damage from power surges caused by lightning or electrical grid issues.

b. Test Surge Protectors:
  Periodically test surge protectors to ensure they are functioning correctly. Replace surge protectors that show signs of wear or are no longer effective.

6. Tree and Vegetation Management:

a. Clear Overhanging Branches:
  Trim overhanging branches that may come into contact with power lines. Preventing contact reduces the risk of power outages and potential electrical hazards.

b. Clear Debris:
  Remove leaves, debris, and vegetation around outdoor electrical components. Keeping areas clear minimizes the risk of fire hazards and ensures accessibility.

7. Record-Keeping:

a. Document Repairs and Changes:

Maintain a record of any electrical repairs, upgrades, or changes made to your home. This documentation is valuable for future reference and potential property transactions.

b. Code Compliance Certificates:

Keep track of code compliance certificates for electrical work. These documents attest to the adherence of work to relevant electrical codes.

8. Upgrade Outdated Systems:

a. Evaluate Electrical Panels:

Assess the capacity and condition of your electrical panel. Consider upgrading to a higher-capacity panel if your current one is outdated or showing signs of wear.

b. Replace Outdated Wiring:

If your home has outdated wiring, consider a comprehensive rewiring project. This enhances safety, reliability, and compliance with modern electrical standards.

Recap: Troubleshooting common electrical issues involves addressing tripped breakers, flickering lights, dead outlets, high energy bills, and electric shocks. Regular maintenance practices include visual inspections, testing and measurements, tightening connections, cleaning and dusting, surge protection, vegetation management, record-keeping, and upgrading outdated systems. These proactive measures contribute to a reliable and safe electrical system in your home.

Chapter 10

Advanced Wiring Projects

Installing a Subpanel

Embarking on advanced wiring projects can significantly enhance the functionality, efficiency, and technological prowess of your home. This chapter delves into two sophisticated projects: installing a subpanel and wiring for smart home systems. By mastering these endeavors, you'll not only expand your electrical expertise but also modernize your living spaces.

1. Assessing the Need:

a. Increased Load Demands:
Consider installing a subpanel when the existing main panel is nearing its capacity. Increased load demands from added appliances or renovations may necessitate a subpanel.

b. Distance from Main Panel:
Evaluate the distance between the main panel and the areas requiring additional circuits. Installing a subpanel can reduce the need for long circuit runs.

2. Selecting the Subpanel:

a. Matching Specifications:

Choose a subpanel that matches the specifications of your main panel. This includes the same voltage (usually 120/240V) and a sufficient number of spaces for circuit breakers.

b. Ampacity Considerations:

Determine the ampacity needed for the subpanel based on the additional circuits it will support. Common residential subpanels range from 60 to 200 amps.

3. Planning Circuit Layout:

a. Identify Circuits:

Plan the circuits that will be connected to the subpanel. This could include dedicated circuits for specific rooms, appliances, or workshops.

b. Load Distribution:

Distribute the load evenly across the subpanel's circuits to prevent overloading. Consider grouping related circuits for better organization.

4. Installation Steps:

a. Turn Off Power:
Prioritize safety by turning off the power at the main panel before beginning the installation. Confirm power is off using a voltage tester.

b. Mounting the Subpanel:
Securely mount the subpanel to a sturdy surface. Ensure proper clearances and accessibility for future maintenance.

c. Connecting Feeders:
Run feeder cables from the main panel to the subpanel. Connect the hot, neutral, and ground wires to their respective terminals in both panels.

d. Installing Circuit Breakers:
Install circuit breakers in the subpanel according to your planned layout. Ensure proper sizing and adherence to electrical codes.

e. Grounding and Bonding:
Establish proper grounding and bonding for the subpanel. Connect the grounding conductor to the ground bar and ensure bonding between neutral and ground.

f. Test and Inspect:
Test the functionality of each circuit in the subpanel and conduct a thorough visual inspection. Confirm compliance with local electrical codes.

5. Expansion and Future-Proofing:

a. Reserve Spaces:
Leave some empty spaces in the subpanel for future expansion. This allows for additional circuits without the need for immediate upgrades.

b. Consider Smart Technologies:
When installing a subpanel, consider incorporating smart technologies for monitoring and control. This future-proofs your electrical infrastructure.

Wiring for Smart Home Systems

1. Planning Smart Home Features:

a. Identify Automation Needs:
Determine the smart home features you want to incorporate, such as lighting control, smart thermostats, security systems, and home automation hubs.

b. Compatibility Considerations:
Ensure that the selected smart devices and systems are compatible with each other and with your chosen home automation platform.

2. Smart Wiring Basics:

a. Run Low-Voltage Wiring:
Plan and run low-voltage wiring to locations where smart devices will be installed. This includes wiring for smart switches, sensors, and other control devices.

b. Centralized Hub:
Consider installing a centralized hub or controller that serves as the brain of your smart home system. This hub connects to your network and facilitates communication between devices.

3. Installation Steps:

a. Wiring for Lighting Control:
Replace traditional switches with smart switches that can be controlled remotely. Ensure proper wiring to accommodate the additional features of smart switches.

b. Thermostat Wiring:
If installing a smart thermostat, follow the manufacturer's instructions for wiring. This may involve connecting to the HVAC system and providing power.

c. Security System Wiring:
If incorporating a smart security system, run wiring for cameras, sensors, and control panels. Ensure secure and discreet wiring to maintain aesthetics.

d. Network Infrastructure:
Ensure robust Wi-Fi or Ethernet connectivity throughout your home. Strong network infrastructure is crucial for the seamless operation of smart devices.

4. Integration and Programming:

a. Pairing Devices:
Pair smart devices with the central hub or controller according to the manufacturer's instructions. Follow proper security protocols for device pairing.

b. Programming Automation:

Set up automation routines and schedules using the control platform. This may involve creating scenes, triggers, and rules for different scenarios.

5. Accessibility and User Interface:

a. User-Friendly Interface:

Opt for user-friendly interfaces to control smart devices. This could include smartphone apps, voice commands, or dedicated control panels.

b. Accessibility Considerations:

Consider the accessibility of smart home features for all household members. Ensure ease of use and compatibility with assistive technologies if needed.

6. Security and Privacy Measures:

a. Secure Network:

Implement robust security measures for your network to prevent unauthorized access to smart devices. Use strong passwords and encryption.

b. Privacy Settings:

Review and adjust privacy settings on smart devices to control data sharing and ensure the protection of personal information.

Recap: Advanced wiring projects involve installing a subpanel to meet increased electrical demands and wiring for smart home systems to integrate automation features. Proper planning, adherence to electrical codes, and consideration of future needs are essential for successful execution. These projects not only enhance functionality but also contribute to the modernization of your home.

Conclusion

As we conclude this comprehensive guide, "Your Guide to Home Electrical Wiring, Outlet and Switch Installs," the journey through the intricate world of electrical projects comes to fruition. This book has been crafted with the intent to empower you, the reader, with the knowledge and confidence to navigate the complexities of home electrical systems.

From unraveling the fundamentals of voltage, current, and resistance to undertaking advanced projects like installing a subpanel and wiring for smart home systems, each chapter has been meticulously designed to provide actionable insights, clear instructions, and a roadmap for success. Whether you are a novice embarking on your first DIY project or a seasoned enthusiast seeking to tackle more advanced endeavors, this guide aims to be your trusted companion.

Throughout these pages, we've not only delved into the technicalities of electrical work but also emphasized the importance of safety, compliance with codes, and the integration of modern technologies. Your home's electrical system is more than just a network of wires and circuits; it's the lifeblood that powers the heartbeat of your living spaces.

As you undertake electrical projects, remember the guiding principles of safety first, meticulous planning, and a commitment to code compliance. Whether you're running new wires, installing fixtures, or embracing smart home technologies, the knowledge gained from these pages serves as a foundation for a secure and efficient electrical environment.

The ability to troubleshoot, maintain, and embark on advanced projects showcased in this guide positions you not merely as a spectator but as an active participant in the functionality and evolution of your home. Electrical work, once seen as a daunting domain, can now be approached with confidence, understanding, and a sense of empowerment.

Your commitment to learning and applying the principles outlined here not only enhances the safety and efficiency of your home but also lays the groundwork for potential future innovations and advancements. As you embark on your electrical journey, may this guide be a source of inspiration, knowledge, and encouragement.

Remember, electrical work is not just about wires and circuits; it's about creating a home environment that is safe, efficient, and attuned to your needs. Whether you're flipping a switch, enjoying the glow of a well-installed light fixture, or marveling at the convenience of a smart home system, let each electrical project be a testament to your newfound expertise and the journey you've undertaken.

Thank you for entrusting us with a part of your learning journey. As you apply the principles learned here, may your home's electrical system shine brightly, reflecting not only the power of electricity but also the power of your newfound knowledge and skills.

Here's to a future filled with well-lit rooms, smart technologies at your fingertips, and the satisfaction of a job well done. May your electrical endeavors be safe, successful, and a source of pride in the place you call home.